SPACECRAFT

Written by Angela Royston

•

HOW TO USE THIS BOOK

Complete the pages by finding the sticker that fits in each outline. Then read all about the different vehicles.

Funfax™ is an imprint of DK Publishing, Inc.
95 Madison Avenue, New York, NY 10016
Copyright © 1999 Funfax Ltd, Woodbridge, Suffolk, IP12 1AN, England.
All rights reserved.

First published in Great Britain in 1999 by
Funfax Ltd, Woodbridge, Suffolk, IP12 1AN.

Going to the Moon

Only twelve astronauts have ever landed on the Moon. Rockets blasted them into space and computers guided their spacecraft down to the surface—a rocky world with no plants or air.

U.S. flag
The astronauts left this flag on the Moon. It is held out by metal arms because there is no wind to blow it.

Spacesuit
A spacesuit protects the astronaut and gives him oxygen to breathe.

Vostok 1
Vostok 1 carried the first astronaut into space. With Yuri Gagarin strapped inside, it orbited Earth once.

Saturn V
This rocket launched Apollo spacecraft to the Moon. It was as tall as a skyscraper and over half of it contained fuel.

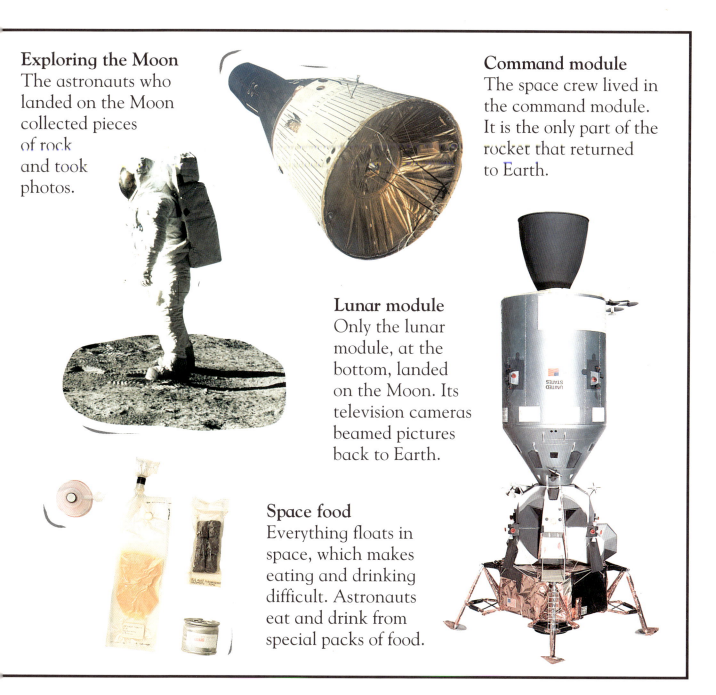

Exploring the Moon
The astronauts who landed on the Moon collected pieces of rock and took photos.

Command module
The space crew lived in the command module. It is the only part of the rocket that returned to Earth.

Lunar module
Only the lunar module, at the bottom, landed on the Moon. Its television cameras beamed pictures back to Earth.

Space food
Everything floats in space, which makes eating and drinking difficult. Astronauts eat and drink from special packs of food.

Probes and telescopes

Probes and telescopes explore space. A probe is guided by computers. It sends pictures and information about other planets back to Earth. It has no crew.

Venera 9
Venera 9 landed on the surface of the planet Venus. It sent back the first photo of another planet.

Viking lander
A Viking lander was the first probe to explore Mars. It found a cold, lifeless planet

Mariner 10
Mariner 10 is the only probe to have visited Mercury, the closest planet to the Sun.

Hubble Space Telescope
The Hubble Telescope looks deep into space. It can see much farther than any other telescope.

Satellites

Hundreds of satellites circle the Earth. They each do a special job. Some record the weather, others beam signals around the world.

Collecting information
Some satellites scan the Earth from space. They use cameras and radar to collect particular information.

Navstar
Signals from Navstar tell ships, planes, and even hikers exactly where they are.

Telstar
Telstar was the first satellite to send television pictures around the world.

Tethered satellite system
A tethered satellite is joined to a spacecraft by a strong, thin cable.

Space shuttle

Spacecraft cost a lot of money and most can only be used once. But the space shuttle can be used many times. It is like a space taxi.

Shuttle at work
The orbiter carries new satellites in its hold. Once in space, the hold doors open.

The orbiter
The main part of the space shuttle is called the orbiter. It looks like a plane.

Launching
Two rockets and a huge fuel tank launch the orbiter into space. The rockets parachute back to Earth and are used again.

Hitching a ride
The orbiter cannot fly through the air like other planes. Instead, it hitches a ride on the back of a jumbo jet.

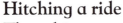

Earth from space
Astronauts on the orbiter get a great view of the Earth from space.

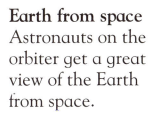

Mini spacecraft
This amazing machine is attached to a spacesuit. It flies the astronaut around outside the spacecraft.

Space visit
The orbiter meets up with the Russian space station, Mir. Mir astronauts can travel back to Earth in it.

Landing
A space shuttle carries astronauts in orbit around the Earth. It glides back to land without engines.

Space stations

A space station is a place where astronauts live and work, sometimes for months at a time. They carry out experiments in space.

Mir
Some astronauts have spent up to a year on Mir to see how long they can live in space.

The International Space Station
New space stations, like this one, have parts added bit by bit.

Skylab
Only three crews worked in Skylab. Today, the U.S. uses space shuttles as space labs instead.

Salyut
The first space stations were launched by the Russians and were called Salyut. This is one of them.